K

Math
Triumphs

Beginning Skills and Concepts

Authors
Basich Whitney • Brown • Dawson • Gonsalves • Silbey • Vielhaber

Mc Graw Hill **Macmillan/McGraw-Hill**
Glencoe

Photo Credits

The *McGraw·Hill* Companies

Macmillan/McGraw-Hill
Glencoe

Send all inquiries to:
Macmillan/McGraw-Hill • Glencoe/McGraw-Hill
8787 Orion Place
Columbus, OH 43240-4027

ISBN: 978-0-07-888193-0
MHID: 0-07-888193-5

Printed in the United States of America.

3 4 5 6 7 8 9 10 066 16 15 14 13 12 11 10 09

Math Triumphs
Grade K

Math Triumphs

Authors and Consultants

CONSULTING AUTHORS

Frances Basich Whitney
Project Director, Mathematics K–12
Santa Cruz County Office of Education
Capitola, California

Kathleen M. Brown
Math Curriculum Staff Developer
Washington Middle School
Long Beach, California

Dixie Dawson
Math Curriculum Leader
Long Beach Unified
Long Beach, California

Philip Gonsalves
Mathematics Coordinator
Alameda County Office of Education
Hayward, California

Robyn Silbey
Math Specialist
Montgomery County Public Schools
Gaithersburg, Maryland

Kathy Vielhaber
Mathematics Consultant
St. Louis, Missouri

CONTRIBUTING AUTHORS

Viken Hovsepian
Professor of Mathematics
Rio Hondo College
Whittier, California

FOLDABLES Study Organizer **Dinah Zike**
Educational Consultant
Dinah-Might Activities, Inc.
San Antonio, Texas

CONSULTANTS

Assessment

Donna M. Kopenski, Ed.D.
Math Coordinator K–5
City Heights Educational Collaborative
San Diego, California

Instructional Planning and Support

Beatrice Luchin
Mathematics Consultant
League City, Texas

ELL Support and Vocabulary

ReLeah Cossett Lent
Author/Educational Consultant
Alford, Florida

Reviewers

Each person reviewed at least two chapters of the Student Study Guide, providing feedback and suggestions for improving the effectiveness of the mathematics instruction.

Dana M. Addis
Teacher Leader
Dearborn Public Schools
Dearborn, MI

Renee M. Blanchard
Elementary Math Facilitator
Erie School District
Erie, PA

Jeanette Collins Cantrell
5th and 6th Grade Math Teacher
W. R. Castle Memorial Elementary
Wittensville, KY

Helen L. Cheek
K-5 Math Specialist
Durham Public Schools
Durham, NC

Mercy Cosper
1st Grade Teacher
Pershing Park Elementary
Killeen, TX

Bonnie H. Ennis
Math Coordinator
Wicomico County Public Schools
Salisbury, MD

Sheila A. Evans
Instructional Support Teacher - Math
Glenmount Elementary/Middle School
Baltimore, MD

Lisa B. Golub
Curriculum Resource Teacher
Millennia Elementary
Orlando, FL

Donna Hagan
Program Specialist - Special Programs
    Department
Weatherford ISD
Weatherford, TX

Russell Hinson
Teacher
Belleview Elementary
Rock Hill, SC

Tania Shepherd Holbrook
Teacher
Central Elementary School
Paintsville, KY

Stephanie J. Howard
3rd Grade Teacher
Preston Smith Elementary
Lubbock, TX

Rhonda T. Inskeep
Math Support Teacher
Stevens Forest Elementary School
Columbia, MD

Albert Gregory Knights
Teacher/4th Grade/Math Lead Teacher
Cornelius Elementary
Houston, TX

Barbara Langley
Math/Science Coach
Poinciana Elementary School
Kissimmee, FL

David Ennis McBroom
Math/Science Facilitator
John Motley Morehead Elementary
Charlotte, NC

Jan Mercer, MA; NBCT
K-5 Math Lab Facilitator
Meadow Woods Elementary
Orlando, FL

Rosalind R. Mohamed
Instructional Support Teacher - Math
Furley Elementary School
Baltimore, MD

Patricia Penafiel
Teacher
Phyllis Miller Elementary
Miami, FL

Lindsey R. Petlak
2nd Grade Instructor
Prairieview Elementary School
Hainesville, IL

Lana A. Prichard
District Math Resource Teacher K-8
Lawrence Co. School District
Louisa, KY

Stacy L. Riggle
3rd Grade Spanish Magnet Teacher
Phillips Elementary
Pittsburgh, PA

Wendy Scheleur
5th Grade Teacher
Piney Orchard Elementary
Odenton, MD

Stacey L. Shapiro
Teacher
Zilker Elementary
Austin, TX

Kim Wilkerson Smith
4th Grade Teacher
Casey Elementary School
Austin, TX

Wyolonda M. Smith, NBCT
4th Grade Teacher
Pilot Elementary School
Greensboro, NC

Kristen M. Stone
3rd Grade Teacher
Tanglewood Elementary
Lumberton, NC

Jamie M. Williams
Math Specialist
New York Mills Union Free School District
New York Mills, NY

Contents

CHAPTER 2
Compare and Order Whole Numbers

Contents

CHAPTER 4 Introduction to Subtraction

Contents

CHAPTER 5 Two-Dimensional Figures

CHAPTER 6 Three-Dimensional Figures

Contents

CHAPTER 8 Sort Objects by Attributes

Full **Empty**

Contents

CHAPTER
10 **Patterns**

Represent Whole Numbers

Home Connection

English

Dear Family,

Today, in **Chapter 1, Represent Whole Numbers,** I started learning about the numbers 0 through 10. I will learn to count and read the numbers 0 through 10.

Love, _____

Spanish

Querida Familia:

Hoy, en **Capítulo 1, Representar números enteros,** empecé a aprender los números de 0 a 10. Aprenderé a contar y leer los números de 0 a 10.

Con cariño, _____

Help at Home

Help your child match numbers to groups of 0 to 10 objects. Write the numbers 0 through 10 each on a separate sheet of paper. Have your child place the correct amount of small objects on each paper and then read each number.

Math Online ▷ Take the chapter Quick Check quiz at macmillanmh.com.

Ayuda en casa

Ayude a su hijo a relacionar los números con grupos de 0 a 10 objetos. Escriba los números 0 a 10 cada uno en hoja separada. Pídale a su hijo que coloque la cantidad correcta de objetos sobre cada una de esas hojas y luego que lea cada número.

Name _____

Get Ready

①

②

③

④

Directions:
1. Circle the puppy.
2. Draw an X on the fish.

3. Color all of the squares.
4. Draw a circle for each triangle.

Name _____

Count Objects 0 to 5

 1

2

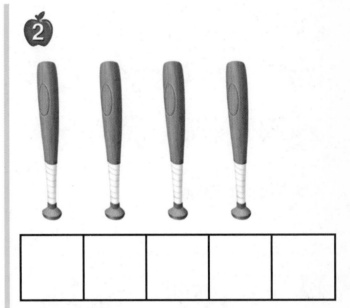

3

Directions:

1–2. Count the objects. Draw a line from each object to one box. Color the boxes to show how many.

3. Draw Xs on 3 footballs.
4. Draw Xs on 4 baseball mitts.

Name _____

Count Objects 6 to 10

①

②

③

④

Directions:

1. Circle 8 trucks.
2. Circle 6 boats.

3. Draw Xs on 7 cars.
4. Draw Xs on 9 boats.

4 four

Name _____

Count Forward

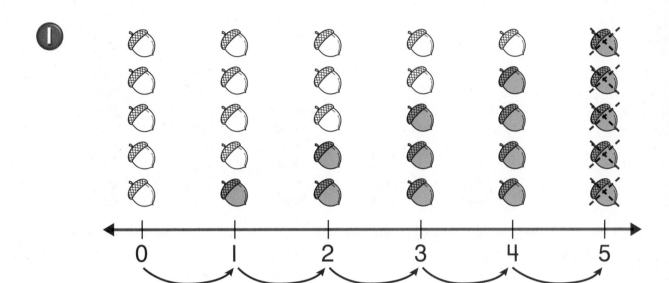

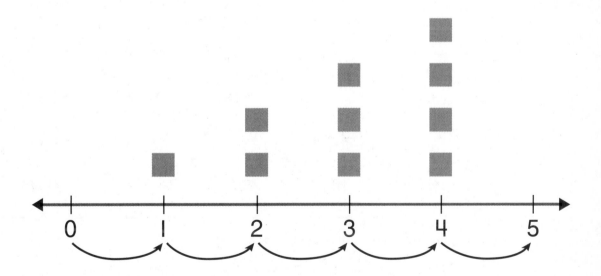

Directions:

1. Count forward to find the group of 5 colored acorns. Draw Xs on the group of 5.

2. Draw the number of squares that comes next.

Name _____

Count Backward

1

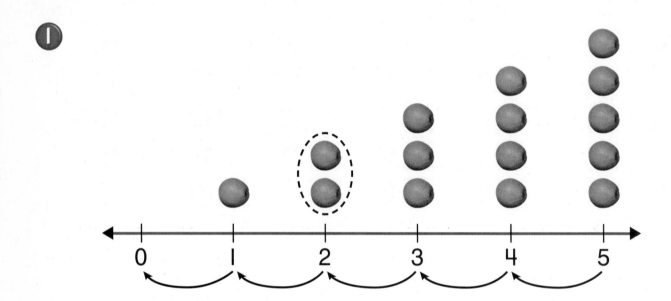

2

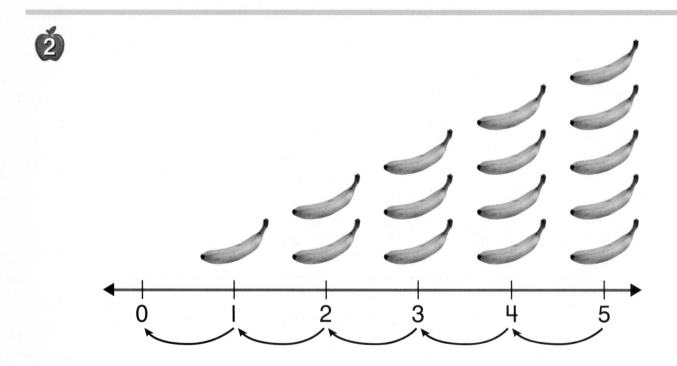

Directions:

1. Start at 5 and count backward. Circle the group of 2.

2. Start at 5 and count backward. Draw an X on the group of 1.

Name _____

Progress Check 1

 1

2

3

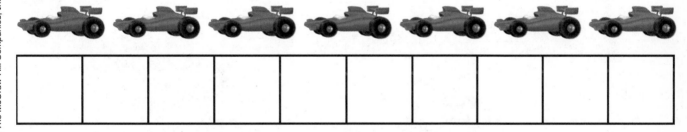

4

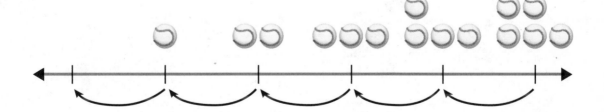

Directions:

1. Count the objects. Draw an X on the group of 2.

2. Circle the group of 4. Draw an X on the group of 3.

3. Count the cars. Shade in the same number of boxes.

4. Start at 5 and count backward. Circle the picture that shows 2.

Name

Replay

Directions:

Step 1: Count the animals on the left-hand side of the page.

Step 2: Draw a line from each animal group to the group of objects with the same number.

Name _____

Numbers 0 and 1

1

 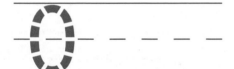

2

3

4

Directions:

1. Here are the things you might see at a party. How many piñatas are there? Trace the number. Write the number two more times.

2. How many gifts are at the party? Trace the number. Write the number two more times.

3. How many violins are there? Trace the number. Write the number three more times.

4. How many horns are there? Trace the number. Write the number three more times.

Name _____

Numbers 2 and 3

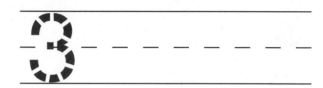

Directions:

1. Color two leaves. Trace the number 2. Write it three more times.
2. Draw three flowers. Trace the number 3. Write it three more times.
3. Count the trees. Trace the number 2. Write it three more times.
4. Count the logs. Trace the number 3. Write it three more times.

Name _____

Numbers 4 and 5

 ①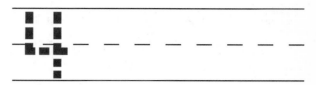

 ②

 ③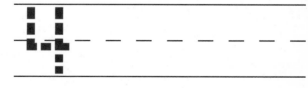

 ④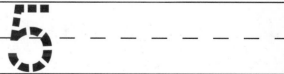

Directions:

1. Count the fish. Trace the number 4. Write it three more times.
2. Draw 5 beach balls. Trace the number 5. Write it three more times.
3. Circle the group of 4. Trace the number 4. Write it three more times.
4. Circle the group of 5. Trace the number 5. Write it three more times.

Name _____

Numbers 6 to 10

1 6

2  7

3  8

4 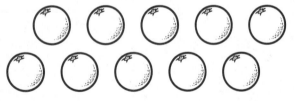 9

5 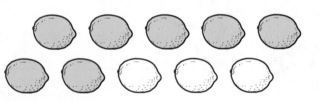 10

Directions:

1. Circle the group of 6. Trace the number 6. Write it 3 more times.
2. Draw 7 bananas. Trace the number 7. Write it 3 more times.
3. Color 8 strawberries. Trace the number 8. Write it 3 more times.
4. Color 9 oranges. Trace the number 9. Write it 3 more times.
5. Color 3 more lemons to make 10. Trace the number 10. Write it 3 more times.

Name _____

Progress Check 2

  _____

  _____

  _____

    _____

Directions:

1. Circle the group of 4. Write the number 4.
2. Count the hammers. Write the number.
3. How many cans of red paint are shown in the picture? Write the number.
4. Draw Xs on 8 paint brushes. Write the number.

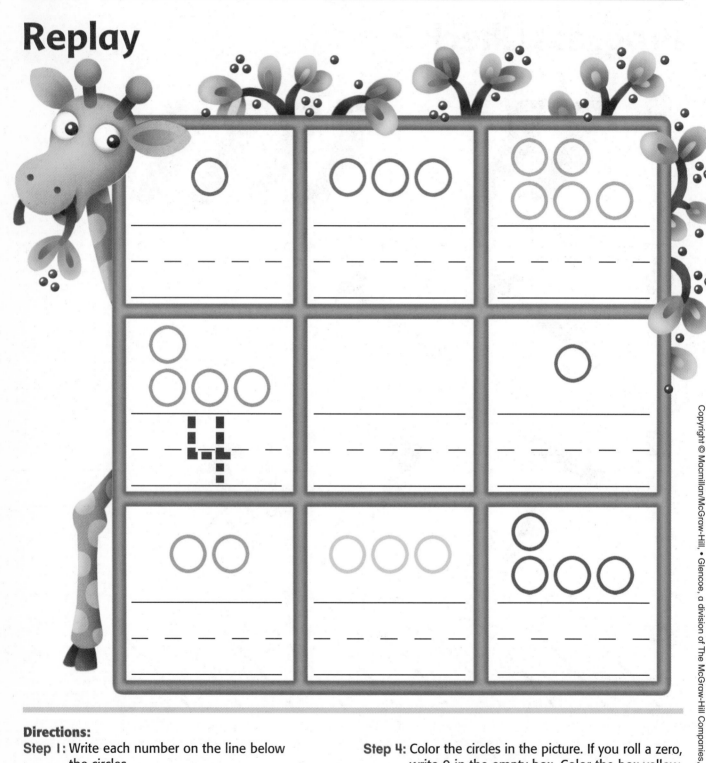

Name _____

Replay

Directions:

Step 1: Write each number on the line below the circles.

Step 2: Roll a number cube labeled 0 to 5.

Step 3: Find a picture with a number of circles that matches the number on the cube.

Step 4: Color the circles in the picture. If you roll a zero, write 0 in the empty box. Color the box yellow.

Step 5: Roll the number cube five times.

Name

Review

 ①

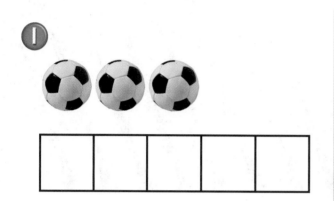

②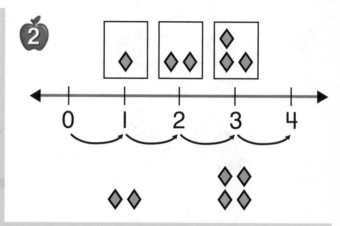

③ _____
_____
_____

 _____
_____

 ④

_____
_____
_____

Directions:

1. Count the soccer balls. Draw a line from each soccer ball to one box. Color the boxes to show how many soccer balls.

2. Circle the picture that comes next on the number line.

3. Circle the group of 6. Draw an X on the group of 5. Write each number.

4. How many presents are there? Write the number.

Chapter 1   Review

Name _____

Test

1

2

3

_____

_ _ _ _

_____

4

_____

_ _ _ _

_____

Directions:
1. Circle the group of 4. Draw an X on the group of 8.
2. Start at 5 and count backward. Draw an X on the picture that shows 2.

3. Count the drums. Write the number of drums.
4. How many bells are in the picture? Write the number.

Home Connection

English   **Spanish**

Dear Family,
Today I started **Chapter 2, Represent Whole Numbers.** I will learn about the numbers 6 through 10. I will learn to count and read the numbers 6 through 10.

Love, _____

Querida Familia:
Hoy empecé **Capítulo 2, Representar números enteros.** Estoy aprendiendo los números de 6 a 10. Aprenderé a contar y leer los números de 6 a 10.

Con cariño, _____

Help at Home

Help your child match numbers to groups of 6 to 10 objects. Write the numbers 6 through 10 on separate sheets of paper. Have your child place the correct number of small objects on each paper, then read each number.

Ayuda en casa

Ayude a su hijo a relacionar los números con grupos de 6 a 10 objetos. Escriba los números 6 a 10 en hojas separadas. Pídale a su hijo que coloque la cantidad correcta de objetos sobre cada una de esas hojas y luego, que lea cada número.

Math Online > Take the chapter Quick Check quiz at macmillanmh.com.

Name _____

Get Ready

①

②

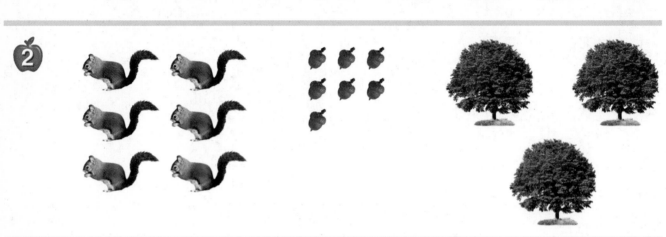

3 | 1 | 2 | 3 | 4 | 5 |

one   two   three   four   five

④

0   1   2   3   ___

Directions:
1. Circle 6 rabbits.
2. Circle the group of 7. Draw an X on the group of 3.
3. Circle the number that tells how many mice.
4. Write the next number.

Name _____

Before and After

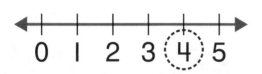 **1**
0 1 2 3 (4) 5

2
0 1 2 3 4 5

3
  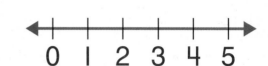

4   _____   6

4

_____   1   2

_____

Directions:

1. Circle the number that comes just before 5.
2. Circle the number that comes just after 2.

3. Write the number that comes just after 4.
4. Write the number that comes just before 1.

Name _____

First, Next, Last

1

2

3

4

1   2   3   4   __ __ __

__ __

Directions:
1. Circle the monkey that is first in line.
2. Circle the fish that is last in line.

3. Find the first turtle. Circle the turtle that is next in line.
4. Write the number that comes next.

Name _____

Second and Third

①

②

③

④

Directions:

1. Circle the second ant.
2. Circle the third bee.

3. Circle the third caterpillar.
4. Circle the second butterfly.

Name _____

Fourth and Fifth

1

2

3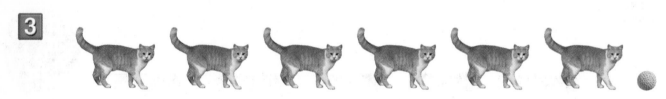

4

Directions:
1. Circle the fifth cat in line.
2. Circle the fifth animal in line.
3. Circle the fourth animal in line.
4. Circle the fourth lion cub in line.

Name _____

Progress Check 1

 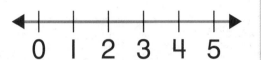   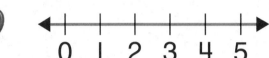

Directions:
1. Circle the number that comes just before 4.
2. Circle the number that comes just after 0.
3. Circle the first rabbit in line. Draw an X on the last rabbit in line.
4. Circle the second cow.
5. Circle the fourth deer.

Name _____

Replay

Directions: Help make a fruit salad.

1. Circle the first apple. Draw a line from the first apple to the bowl.
2. Circle the second orange. Draw a line from the second orange to the bowl.
3. Circle the last bunch of grapes. Draw a line to the bowl from this bunch of grapes.
4. Circle the fourth bunch of bananas. Draw a line from the fourth banana to the bowl.
5. Circle the first bunch of cherries. Draw a line to the bowl from this bunch of cherries.
6. Circle the third pear. Draw a line from the third pear to the bowl.

Name _____

Equal Sets

 1

 2

3

Directions:

1. Circle the group with the same number of animals as the group of lions.
2. Circle the group with the same number of animals as the group of pigs.
3. Draw circles to make a group of circles equal to the group of frogs.

Name _____

Greater Than and Less Than

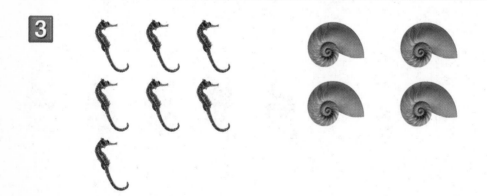

Directions:
1. Circle the group that is less than the other.
2. Circle the group that is greater than the other.

3. Draw an X on the group that is less than the other.
4. Draw an X on the group that is greater than the other.

Name

Growing Number Patterns

1

1   2   3   4   5

2

1   2   3   4   ____

3

4

Directions:

1–2. Look for a pattern. Write the missing number.

3–4. Look for a pattern. Draw a picture in the box for the missing number.

Name

More Number Patterns

1.

8   7   _____   5

2.

9   8   _____   6   5

3.

8   7   _____   5   4   _____   2

Directions:

1–2. Look for a pattern. Write the missing number.

3. Look for a pattern. Write the missing numbers.

Name _____

Progress Check 2

①

②

③

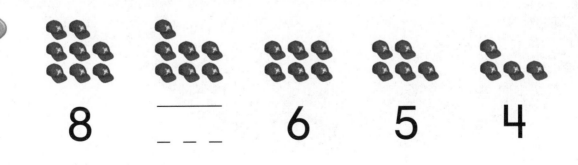

0   1   2   3   ___   5   6

④

8   ___   6   5   4

Directions:
1. Draw circles to make a group of circles equal to the group of hats.
2. Circle the group that is less than the other.
3. Look for a pattern. Write the missing number.
4. Look for a pattern. Write the missing number.

Name _____

Replay

0 1 2 3 4 5 6 7 8 9 10

5	1	2
6	4	8
9	7	10

2 4 9

_____

Directions:
How many apples did the pig eat?
Step 1 In the table, find the numbers greater than 4.
Color them purple.
Step 2 In the table, find the numbers less than 3.
Color them orange.

Step 3 Look at the apples. Circle the apple with the number that is not colored.
Step 4 Write the number.

30 thirty

Name

Review

①

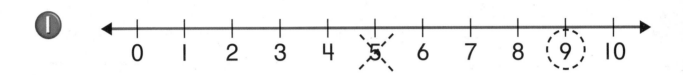

0 1 2 3 4 ~~5~~ 6 7 8 (9) 10

②

3

④

Directions:

1. Circle the number that comes just before 10. Draw an X on the number that comes just after 4.
2. Circle the first rabbit. Draw an X on the last rabbit. Draw a square around the fifth rabbit.
3. Circle the groups of birds with the same number of animals as the group of ducks.
4. Look for a pattern. Draw circles for the number that comes next.

Test

1

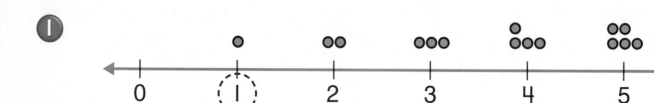

2

3

4

5

6

  4   5   6   ___   8

Directions:

1. Circle the number that comes just before 2.
2. Circle the fifth shape. Draw an X on the second shape.
3. Draw circles to make the sets equal.

4. Circle the group that is less than the other.
5. Circle the group that is greater than the other.
6. Look for a pattern. Write the missing number.

32 thirty-two

Home Connection

English

Dear Family,

Today in **Chapter 3, Introduction to Addition,** I started learning about addition. I will learn to join sets to create groups up to 9.

Love, _____

Spanish

Querida Familia:

Hoy, en el **Capítulo 3, Introducción a la suma,** empecé a aprender la suma. Aprenderé a unir grupos para crear sumas hasta 9.

Con cariño, _____

Help at Home

Help your child understand addition. Model addition at home. If you put 2 cans of soup on a shelf and 3 cans are already there, ask your child, "How many soup cans are there?"

Ayuda en casa

Ayude a su hijo a comprender la suma. Ejemplifique problemas de suma en su casa. Si pone dos latas de sopa en el estante y ya hay otras 3 latas, pregúntele a su hijo: "¿Cuántas latas de sopa hay en total?"

Math Online > Take the chapter Quick Check quiz at macmillanmh.com.

Name _____

Get Ready

①

_____
- - - - ▮ - - - -
_____

②

_____

- - - - - - - - -

_____

③

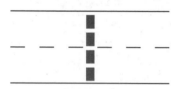

④

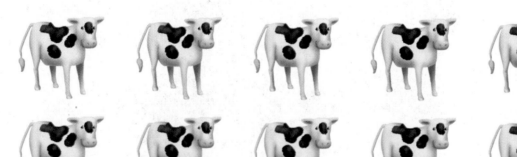

Directions:
1. Write the number of horses.
2. Write the number of pigs.
3. Circle the group of 5.
4. Circle 8 cows.

Chapter 3   Get Ready

_____

_____

Name _____

Sums of 1 and 2

①

②

③

④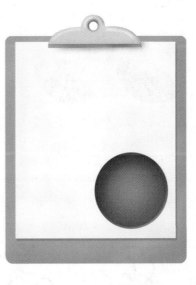

Directions:

1. Alex has 1 red crayon. His teacher gives him 1 blue crayon. How many crayons does Alex have in all?
2. Mave has zero pencils. Then her teacher gives her 1 pencil. How many pencils does Mave have in all?
3. Draw 1 counter on the paper. Write how many counters in all.
4. Draw 2 more counters on the paper. Write how many counters in all.

Name _____

Sums of 3 and 4

①

3

②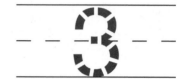

③

④

Directions:

1. Matt has one football. Derek gives him two baseballs. How many balls does Matt have in all?
2. Devon has 3 football helmets. His dad gives him 1 bicycle helmet. How many helmets does he have in all?
3. The ball bin has 2 footballs. Joan puts a tennis ball in the bin. How many balls are in the bin in all?
4. There are 2 bicycles at the bike rack. Then 2 more bicycles are put at the bike rack. How many bicycles are at the rack in all?

Name _____

Sums of 5

(1)

_____

(2)

_____

(3)

_ _ _ _

(4)

_ _ _ _

Directions:

1. Lucy has 4 shoes in her closet. She finds 1 more shoe under her bed. How many shoes does she have in all?
2. Amy has 3 shirts. She has 2 pairs of jeans. How many items does she have in all?
3. Mark has 1 cap. His mom gives him 4 mittens. How many items does he have in all?
4. José has 0 cowboy hats. He buys 5 cowboy hats. How many hats does he have in all?

Name _____

Sums of 6

 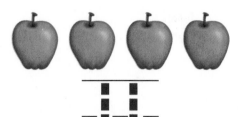 

_____   _____   _____

- - -   - - -   - - -

_____   _____   _____

_____   _____   _____

- - -   - - -   - - -

_____   _____   _____

Directions:

1. How many red apples are there? How many yellow apples are there? How many apples are there in all?

2. The bears eat 3 cherries. Then they eat 3 bananas. How many pieces of fruit did the bears eat in all?

3. Raul's cart has 6 strawberries. There are 0 limes. How many pieces of fruit are on Raul's cart in all?

4. You have 4 lemons. Color the number of lemons you need to have 6 in all.

Name _____

Progress Check 1

   _____

_____

 _____

  _____

_____

8

_____   _____

_____   _____

_____   _____

_____   _____

Directions:
1. Susan has 1 blue counter. Her teacher gives her 1 green counter. How many counters does she have in all?
2. Bob has three shirts. His mom gives him 2 caps. How many items does Bob have in all?
3. Kira has 4 rings. Her friend has 2 rings. How many rings do they have in all?
4. Ms. Jones found 2 backpacks in her class. She found 2 jackets in the gym. How many items did she find in all?

Name _____

Replay

Directions:

Step 1: Color the engine blue. How many cars have you colored in all?

Step 2: Color 2 train cars red. How many cars have you colored in all?

Step 3: Color 1 train car green. How many cars have you colored in all?

Step 4: Color 2 train cars yellow. How many cars have you colored in all?

Name

Sums of 7

3

Directions:
1. Draw more leaves to make a total of 7.
2. Mimi has 4 flowers. She picks 3 more flowers. How many flowers does she have in all?
3. James saw 2 trees. Then he saw 5 more trees. How many trees did he see in all?
4. Marta has 1 flower. Color more flowers to show 7 flowers in all.

Name

Sums of 8

①

1   7   8

②

_____   _____   _____

- - -   - - -   - - -

_____   _____   _____

3

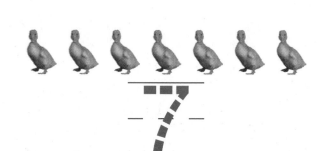

_____

- - -

④

Directions:

1. There is 1 duck in the pond. Then her 7 baby ducks join her. How many ducks are there in all?
2. Samantha saw 4 kittens. She also saw 4 guinea pigs. How many animals did she see in all?
3. Mario has 2 dogs. His neighbor has 6 dogs. How many dogs are there in all?
4. Draw enough fish to have 8 in all.

Name

Sums of 9

Directions:

1. Sam has 3 pieces of chalk and his teacher gives him 6 more pieces of chalk. How many pieces of chalk does Sam have in all?
2. Felipe has 5 pencils. He sharpens 4 more pencils. How many pencils does Felipe have in all?
3. Ms. Jackson has 6 pairs of scissors. Draw Xs to show 9 in all.
4. Mr. Liu has 4 bottles of glue. Color more bottles to show 9 in all.

Name _____

Progress Check 2

1    _____

2

_____   _____   _____

3

_____   _____   _____

Directions:
1. Molly put 1 banana in a bowl. Then she added 7 cherries to the bowl. *How many pieces of fruit are in the bowl in all?*
2. Juan washed 4 red apples. Then he washed 3 yellow apples. *How many apples did Juan wash in all?*
3. Derek bought 1 lemon. Then he bought 8 pears. *How many pieces of fruit did Derek buy in all?*
4. Draw more oranges to make 7 in all.

Name _____

Review

1  _____ _ _ _ _ _____

2

_____ _ _ _ _ _____

3

_____   _____   _____

_ _ _ _   _ _ _ _   _ _ _ _

_____   _____   _____

4

Directions:

1. The team shared 3 basketballs. Then they shared 3 footballs. How many balls did they share in all?

2. A store sells 4 bicycles in one day. The next day the store sells 4 more. How many bicycles were sold in all?

3. Write the number of hats in each group. Write how many hats there are in all.

4. Draw enough baseballs to make a total of 7.

Name _____

Test

1  _____

2

_____   _____   _____

_ _ _ _ _   _ _ _ _ _   _ _ _ _ _

_____   _____   _____

3

_____   _____   _____

_ _ _ _ _   _ _ _ _ _   _ _ _ _ _

_____   _____   _____

4

Directions:

1. Two yellow birds were in the tree. Then 6 blue birds landed in the tree. How many birds are in the tree in all?

2–3. Write how many animals are in each group. Then write how many there are in all.

4. There are 3 fish in the tank. Ana wants 5 fish in the tank. Color the number of fish Ana needs to have 5 fish in the tank.

Home Connection

English

Dear Family,
Today I started **Chapter 4, Introduction to Subtraction.** I am learning about subtraction. I will learn to subtract from numbers up to 9.

Love, _____

Spanish

Querida Familia:
Hoy comencé **Capítulo 4, Introducción a la resta.** Estoy aprendiendo la resta. Aprenderé a restar hasta el número 9.

Con cariño, _____

Help at Home

Help your child understand subtraction. Shake 9 pennies in your hands and drop them on a table. Have your child remove pennies that land tails-side up. Ask how many pennies are left.

Math Online > Take the chapter Quick Check quiz at macmillanmh.com.

Ayuda en casa

Ayude a su hijo a entender la resta. Sacuda 9 monedas de un centavo en su mano y déjelas caer sobre la mesa. Juegue a cara o cruz. Pídale a su hijo que levante los centavos que cayeron en cruz. Pregúntele a su hijo cuántos centavos quedan.

Name _____

Get Ready

1       _____

2 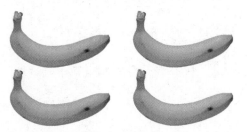      _____

3       _____

4       _____

Directions:

1. Susana's Market sold 4 bananas in the morning. It sold 1 banana in the afternoon. How many bananas did Susana's Market sell in all?
2. Chloe ate 6 cherries for a snack. Later she ate 2 more. How many cherries did she eat in all?
3. Rabi had 3 tomatoes. Then he bought 2 more. How many tomatoes does he have in all?
4. The Good Earth Restaurant used 5 heads of lettuce during lunch. It used 4 heads of lettuce during dinner. How many heads of lettuce did it use in all?

4-1

Name _____

Take Away from 1 and 2

$$- \mathbf{1}$$

_ _ _

_____
_ _
_____

_____
_ _
_____

Directions:

1. There were 2 bicycles parked at school. Tom takes away 1. How many bicycles are left at school?
2. There was 1 car in the parking lot. Then 1 car drove away. How many cars are left?
3. There were 2 trucks delivering food. Then 2 trucks drove away. How many trucks are left?
4. There were 2 people in rowboats. Then 1 boater went back to shore. How many people are left in the boats?

Name _____

Take Away from 3 and 4

1

2      _____

3   _____

4      _____

Directions:

1. Sue had 4 balloons. She gave 1 to her friend. How many balloons does Sue have left?

2. Rosa wrapped 3 presents. She gave 2 of the presents away. How many presents does she have left?

3. Joe had 4 party hats. He gave 2 hats to friends. How many hats does he have left?

4. Pappy's Party Store had 3 piñatas. Luke bought 1. How many piñatas does the store have left?

Name _____

Take Away from 5

 ①

 ②   _ _ _

③   _ _ _

 ④   _ _ _

Directions:

1. There were 5 butterflies on a tree. Then 2 butterflies flew away. How many butterflies are left on the tree?

2. There were 5 ladybugs sitting in the sun. Then 4 ladybugs crawled away. How many ladybugs are left?

3. There were 5 pansies in the garden. Amy put 1 flower in her hair. Draw an X on 1 flower. How many flowers are left in the garden?

4. There were 5 leaves on the branch. Then 3 leaves fell off the branch. Draw an X on 3 leaves. How many leaves are left on the branch?

Name

Take Away from 6

1

2

3

4

Directions:

1. The kicker had 6 footballs to use for practice. He kicked away 2 footballs. How many footballs does he have left?

2. The team had 6 baseballs. They gave 3 to the fans. How many baseballs does the team have left?

3. A store had 6 soccer balls. They sold 1 soccer ball. Draw an X on 1 soccer ball. How many soccer balls are left at the store?

4. There were 6 basketballs on the rack. Marty's teacher took away 6 basketballs. Draw an X on 6 basketballs. How many basketballs are left on the rack?

Name _____

Progress Check 1

 ①   _____

②   _____

 ③  _____

 ④  _____

Directions:

1. Mrs. Ramírez had 2 whistles. She gave 1 to the leader. How many whistles does the teacher have left?

2. Mr. Choi had 3 paint brushes. He gave away 1 paint brush. How many paint brushes does the teacher have left?

3. Mr. Mika had 5 sombreros. He gave 3 sombreros to students to wear. How many sombreros does he have left?

4. The math teacher had 6 books. She gave 4 to her students. Draw an X on 4 books. How many books does Mrs. Switzer have left?

Name _____

Replay

Directions:

1. Jimmy had 6 snowballs. He lost 1. How many does he have left? Color this number white.
2. Rob had 4 carrots. He ate 1. How many does he have left? Color this number orange.
3. There were 3 mittens in the closet. Marisol took 2 of them to wear. How many mittens are left? Color this number red.
4. Emma's snowman has 6 buttons. The wind made 0 buttons fall off. How many buttons are left? Color this number green.

5. There were 4 children playing in the snow. Then 2 children went home. How many children are still playing? Color this number yellow.
6. There were 5 icicles hanging from the tree. Then 1 icicle fell off. How many are left on the tree? Color this number black.
7. Andre's snowman had 2 scarves. Then 2 scarves blew away. How many are left on his snowman? Color this number blue.

Name _____

Take Away from 7

 1

 2

3

 4

Directions:

1. A store had 7 alarm clocks. Sam bought 3 alarm clocks. How many alarm clocks are left at the store?
2. Maria had 7 cuckoo clocks. She gave 6 to her friends. How many cuckoo clocks does Maria have left?
3. Alan had 7 pocket watches. He gave 4 to friends. Draw an X on 4 watches. How many pocket watches does he have left?
4. Henry had 7 watches. He gave 2 to his brothers. Draw an X on 2 watches. How many watches does he have left?

Name _____

Take Away from 8

 1

 2

_____

3

_____

 4

_____

Directions:

1. There were 8 bears sleeping. Then 3 bears woke up. How many bears are left sleeping?
2. There were 8 dogs barking. Then 2 dogs stopped barking. How many dogs are left barking?
3. There were 8 horses in the barn. Then 3 horses went outside. Draw an X on 3 horses. How many horses are left in the barn?
4. There were 8 cats in the window. Then 4 cats ran away. Draw an X on 4 cats. How many cats are left in the window?

Name _____

Take Away from 9

 1

 2

3

 4

Directions:

1. There were 9 paintbrushes on the shelf. Tom used 4 paintbrushes. How many are left on the shelf?

2. There were 9 crayons in the box. Wendy took 7 crayons out of the box. How many crayons are left in the box?

3. There were 9 pencils on the desk. Carla put 3 pencils away. Draw an X on three pencils. How many pencils are left on the desk?

4. There were 9 glue sticks. The class used 8 glue sticks. Draw an X on 8 glue sticks. How many glue sticks are left?

Name _____

Progress Check 2

1

2

3

4

Directions:

1. Dan had 7 apples. He ate 2 apples. How many apples are left?
2. The monkeys had 8 bananas. They ate 6 bananas. How many bananas are left?
3. Rosa had 8 cherries. She ate 3 cherries. Draw an X on 3 cherries. How many cherries are left?
4. The bears had 9 oranges. They ate 5 oranges. Draw an X on 5 oranges. How many oranges are left?

Name _____

Review

 1

_____

2

_____

3

_____

 4

_____

Directions:

1. The clown had 4 hats. He lost 1 hat. How many hats does he have left?

2. There were 5 dirty shirts. Molly washed 3 shirts. How many dirty shirts are left?

3. The store had 7 pairs of pants. The store sold 5 pairs of pants. Draw an X on five pairs of pants. How many pairs of pants are left?

4. There were 9 shoes on the floor. Madison put away 6 shoes. Draw an X on 6 shoes. How many shoes are left on the floor?

<Name>

Test

①

_____
- - - - -
_____

②

_____
- - - - -
_____

3

_____
- - - - -
_____

④

_____
- - - - -
_____

Directions:

1. There were 3 cows in the barn. Then 2 cows left the barn. How many cows are still in the barn?
2. There were 6 frogs in the pond. Then 4 frogs hopped out. How many frogs are left in the pond?
3. There were 7 cats playing with balls. Then 4 cats went to sleep. Draw an X on 4 cats. How many cats are left playing?
4. Sara was walking 8 dogs. She took 5 dogs home. Draw an X on 5 dogs. How many dogs are left walking?

60 sixty

CHAPTER 5 Two-Dimensional Figures

Home Connection

English

Dear Family,
Today in **Chapter 5, Two-Dimensional Figures,** I started to learn about two-dimensional figures. I will learn about circles, triangles, rectangles, and squares. I will also learn to build my own two-dimensional figures.

Love, _____

Spanish

Querida Familia,
Hoy en **Capítulo 5, Figuras bidimensionales,** comencé a aprender figures bidimensionales. Aprenderé de circulos, triángulos, rectángulos y cuadrados. También prenderé a construir mis propias figures bidimensionales.

Con cariño, _____

Help at Home

You can work with your child to recognize two-dimensional figures. Have your child tell you the shape of road signs while you are outside.

Ayuda en casa

Puede trabajar con su hijo para reconocer figures bidimensionales. Pídale que le diga que forma tienen las señales de tránsito cuando están en la calle.

 Take the chapter Quick Check quiz at macmillanmh.com.

Name

Get Ready

 1

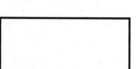

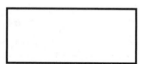

 2

 3

 4

Directions:

1. Color the shapes that are the same.
2. Circle the shapes that are the same.

3–4. Draw an X on the shape that is different.

Name _____

Open or Closed Figures

  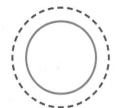

  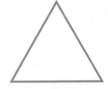

Directions:
1–2. Circle the closed figure.
3–4. Draw an X on the open figure.

5. Draw a closed figure.

Name

Curved or Straight

①

② D

③

④

Directions:
1. Circle the straight line.
2. Trace the straight line with red. Trace the curved line with blue.

3–4. Draw an X on the curved lines.

Name _____

Circles

 1

 2

   _____
_ _ _ _ _ _
_____

3

 4

Directions:
1. Draw an X on the circle.
2. Write the number of sides.

3. Draw an X on the circles.
4. Draw an X on the buttons that are circles.

Name

Triangles

1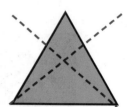

2 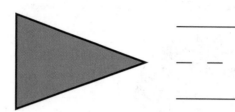 _____
_ _ _ _ _ _
_____

3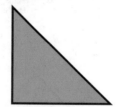

4

Directions:
1. Draw an X on the triangle.
2. Write the number of sides.
3. Draw an X on the figure with 3 corners.
4. Draw an X on the sign that is a triangle.

Name _____

Progress Check 1

1

2

3

4  _ _ _ _  _ _ _ _

5

Directions:
1. Draw an X on the open figure.
2. Circle the straight lines.
3. Draw an X on the circles.

4. Write the number of corners for each figure.
5. Circle the figures with 3 sides.

Name _____

Replay

Start

Finish

Directions:
1. Color all the triangles green.

2. Follow the triangles on the path through the maze.

Name _____

Rectangles

 ①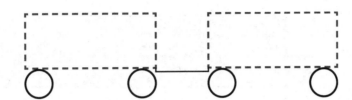

 ②

 ③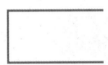

 ④

_____

- - - - -

_____

 ⑤

Directions:
1. Trace the rectangles to make a train.
2–3. Draw an X on the rectangle.

4. Write the number of sides.
5. Draw an X on the shapes with 4 corners.

Name _____

Squares

 1

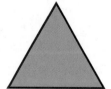

 2

3

 4

Directions:
1. Trace the squares to make a picture of a truck.
2. Draw an X on the square.
3. Write the number of corners on the game board.
4. Draw an X on the square.

Chapter 5   Lesson 6

Name _____

Create Two-Dimensional Figures

 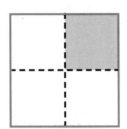

 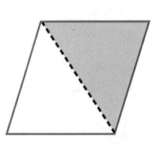

 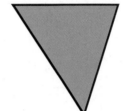 

 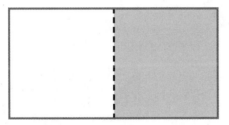 

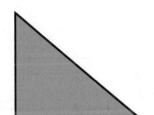  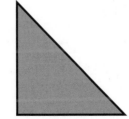 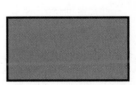

Directions:

1. Draw an X on the figure that is used to make the square.

2. Draw an X on the figure that is used to make the blue figure.

3. Draw an X on the figure that is used to make the rectangle.

4. Draw an X on the figure that is used to make the square.

Name _____

Progress Check 2

 1

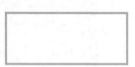

 2

3

 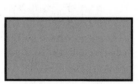

 4

Directions:
1. Draw an X on each figure with 4 corners.
2. Draw an X on each square.
3. Write the number of sides for each figure.

4. What shape would you make if you put the two triangles together? Draw an X on the shape.

Name _____

Review

①

②

③

④

⑤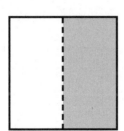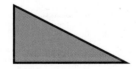

Copyright © Macmillan/McGraw-Hill, • Glencoe, a division of The McGraw-Hill Companies, Inc.

Directions:
1. Draw an X on the open figure.
2. Circle the curved lines.
3. Write the number of sides.

4. Draw an X on each figure with 4 corners.
5. Draw an X on the figure that is made by folding the square in half.

Name _____

Test

 1

2

3

4

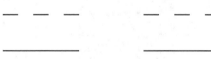

5

6

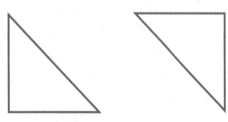

Directions:
1. Circle the closed figure.
2. Draw an X on the straight lines.
3. Circle the figures with 4 corners.

4. Write the number of sides for each figure.
5–6. Draw the shape you would make if you put the two figures together.

74 seventy-four

Chapter 5   Test

Home Connection

English

Spanish

Dear Family,

Today I started **Chapter 6, Three-Dimensional Figures.** I will learn about three-dimensional figures. I will learn how spheres, cylinders, prisms, and cubes move. I will build my own figures.

Love, _____

Querida Familia:

Hoy empecé **Capítulo 6, Figuras tridimensionales.** Aprenderé sobre figuras tridimensionales. Aprenderé cómo se mueven las esferas, los cilindros, prismas y cubos. Crearé mis propias figuras.

Con cariño, _____

Help at Home

Help your child identify shapes. Collect objects of different shapes. Have your child name the three-dimensional shape of each object, such as cube, sphere, rectangular prism, or cylinder.

Ayuda en casa

Ayude a su hijo a identificar las formas. Junte objetos de diferentes formas. Pídale a su hijo que nombre las figuras tridimensionales de cada objeto, como por ejemplo, cubo, esfera, prisma rectangular o cilindro.

 **Math Online** Take the chapter Quick Check quiz at macmillanmh.com.

Name _____

Get Ready

①

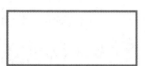

②

③

④

Directions:

1. Color the square.
2. Circle the object that has straight sides.
3. Circle the object that looks like a ball.
4. Draw an X on the object that looks like a square.

Name _____

Introduce Three-Dimensional Figures

①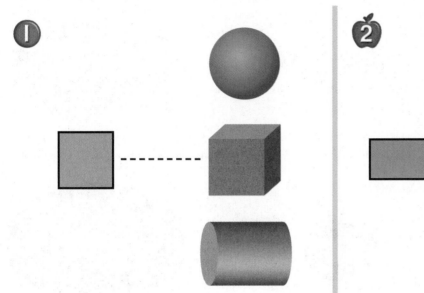

②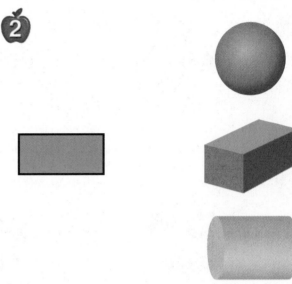

③

④

Directions:

1. Draw a line from the square to the object that has a square as one of its faces.

2. Draw a line from the rectangle to the object that has a rectangle as one of its faces.

3. Draw an X on the object that has a circle as one of its faces.

4. Draw a line from the square to the object that has a square as one of its faces.

Name _____

Roll and Stack

 1

 2

 3

 4

Directions:
1. Circle the object that can be rolled.
2. Circle the object that can be stacked in a tower.
3. Draw an X on the object that can be rolled.
4. Draw an X on the object that can be rolled and stacked.

Chapter 6   Lesson 2

Name _____

Spheres

1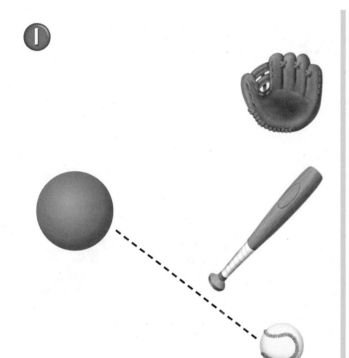

2

3

4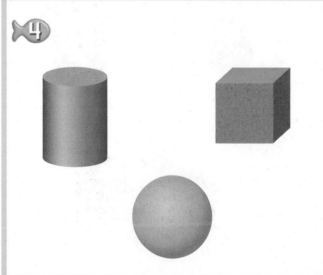

Directions:

1. Draw a line from the green sphere to the other sphere.

2. Draw a line from the purple sphere to the other sphere.

3–4. Draw an X on the sphere.

Name

Cylinders

①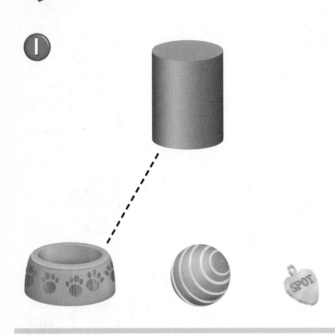

②

③

④

Directions:

1. Draw a line from the blue cylinder to the other cylinder.

2. Draw a line from the orange cylinder to the other cylinder.

3–4. Draw an X on the cylinder.

Name _____

Progress Check 1

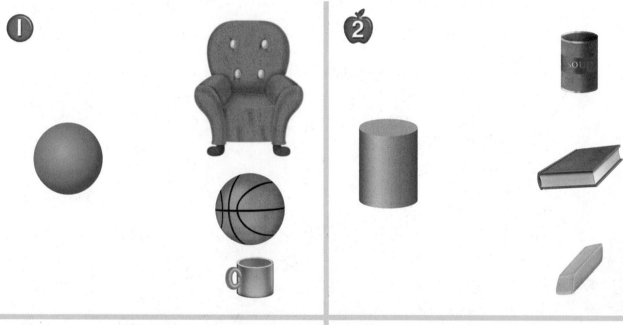

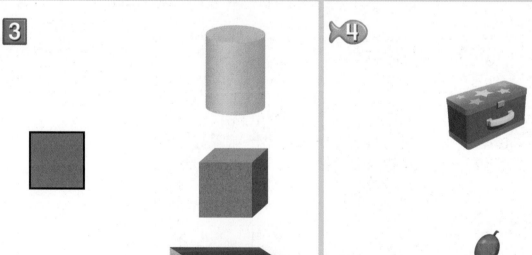

Directions:

1. Draw a line from the green sphere to the other sphere.

2. Draw a line from the blue cylinder to the other cylinder.

3. Draw a line from the square to the figure that has a square as one of its faces.

4. Circle the object that can be stacked.

Name _____

Replay

Directions:
1. Draw an X on objects that can be stacked.
2. Circle the object that has a square as one of its faces.
3. Color the sphere orange.
4. Color the cylinder green.

Name _____

Rectangular Prisms

 1

 - - - -

 2

3

 4

Directions:

1. Draw a line from the blue rectangular prism to the other rectangular prism.

2. Draw a line from the green rectangular prism to the other rectangular prism.

3–4. Draw an X on the rectangular prism.

Name

Cubes

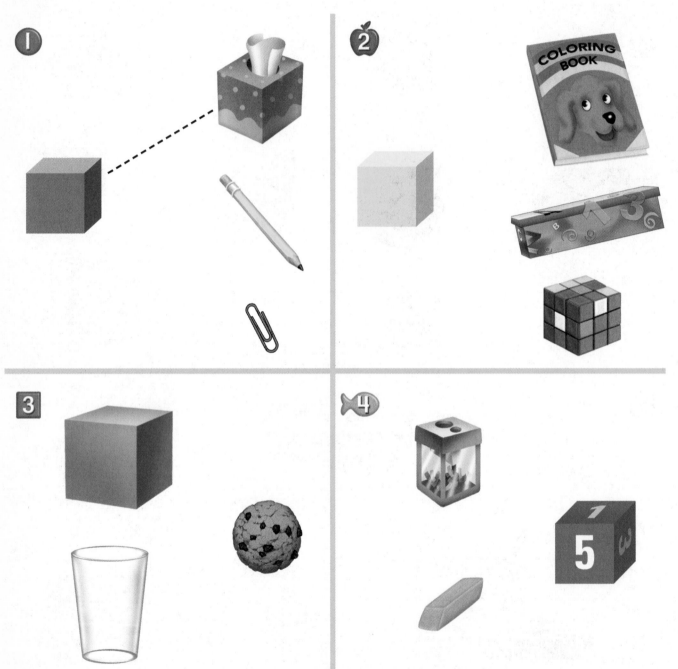

Directions:

1. Draw a line from the blue cube to the other cube.

2. Draw a line from the yellow cube to the other cube.

3–4. Draw an X on the cube.

Name _____

Create Three-Dimensional Figures

1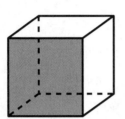

2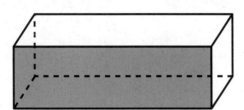

3

4

Directions:

1. Draw an X on the shape that is a face of the cube.
2. Draw an X on the shape that is a face of the rectangular prism.
3. Draw an X on the shapes that make a cylinder.
4. Draw an X on the shape that is a face on the box.

Name _____

Progress Check 2

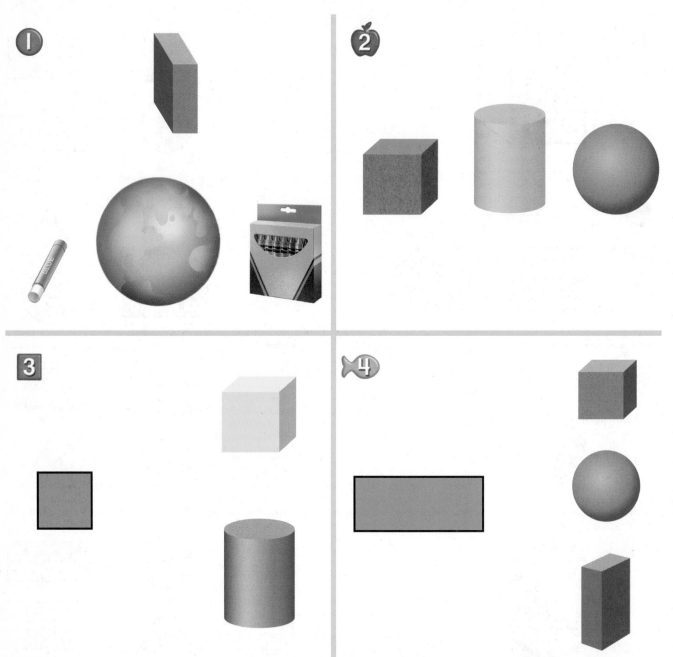

①

②

③

④

Directions:

1. Draw a line from the blue rectangular prism to the other rectangular prism.
2. Circle the cube.

3. Draw a line from the square to the object that has a square as one of its faces.
4. Draw a line from the rectangle to the objects that have a face that is a rectangle.

Name _____

Review

①

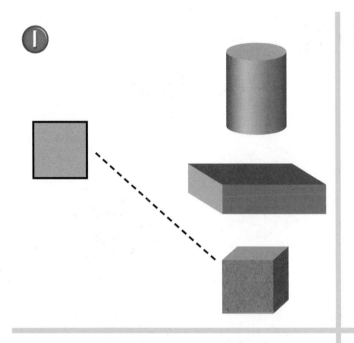

②

③

④

Directions:

1. Draw a line from the square to the object that has a square as one of its faces.
2. Circle the object that can be rolled and stacked.
3. Draw an X on the sphere.
4. Circle the cube.

Name _____

Chapter Test

Copyright © Macmillan/McGraw-Hill, a division of The McGraw-Hill Companies, Inc.

Directions:

1. Draw a line from the red rectangle to the object that has a rectangle as one of its faces.
2. Circle the object that can be rolled.
3. Draw an X on the cylinder.
4. Circle the rectangular prism.

Home Connection

English

Dear Family,

Today, in **Chapter 7, Space and Position,** I started learning about an object's place in space. I will learn words such as "above," "below," "left," and "right." I will use an object's place in space to solve puzzles.

Love, _____

Spanish

Querida Familia:

Hoy en **Capítulo 7, Espacio y ubicación,** empecé a aprender sobre el lugar que ocupa un objeto en el espacio. Aprenderé palabras como "encima," "debajo," "izquierda," y "derecha." Utilizaré el lugar de un objeto en el espacio para resolver acertijos.

Con cariño, _____

Help at Home

Help your child identify an object's place in space. In the kitchen, have your child describe various objects' space and location in relation to other items.

Ayuda en casa

Ayude a su hijo a identificar el lugar que ocupa un objeto en el espacio. Pídale a su hijo que describa el espacio que ocupa y la ubicación de varios objetos en su cocina en relación con otros objetos.

Math Online Take the chapter Quick Check quiz at macmillanmh.com.

Name _____

Get Ready

  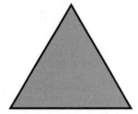 

3

4

Directions:
1. Circle the first ball.
2. Circle the last ball.
3. Draw an X on the triangle.
4. Draw a square.

Name _____

Before or After

 ①

②

 ③

④

Directions:

1. Circle the marble before the gold marble.
2. Circle the marble before the red marble.

3. Draw an X on the marble after the green marble.
4. Draw an X on the marble after the blue marble.

Name

Above or Below

1

2

3

4

Directions:

1. Draw an X above the gift.
2. Draw an X below the piñata.
3. Circle the balloon below the hat.
4. Circle the balloon above the noise-maker.

Chapter 7 Lesson 2

Name _____

Top, Middle, or Bottom

1

2

3

4

Directions:
1. Circle the fruit in the middle.
2. Circle the fruit on top.

3. Circle the fruit on the bottom.
4. Draw a circle on the top.
   Draw an X on the bottom.

Name _____

Left or Right

1

 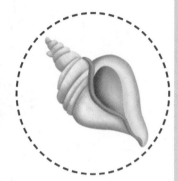

2

3

4

Directions:
1. Circle the shell on the right.
2. Circle the fish on the left.

3. Draw a circle to the left of the starfish.
4. Draw a square to the right of the crab.
   Draw a triangle to the left of the crab.

Name _____

Progress Check 1

Directions:
1. Circle the toy after the jump rope.
2. Circle the toy before the ball.
3. Draw an X below the plane.
4. Draw an X above the truck.
5. Circle the toy car on top.
6. Circle the toy on the left.

Name _____

Replay

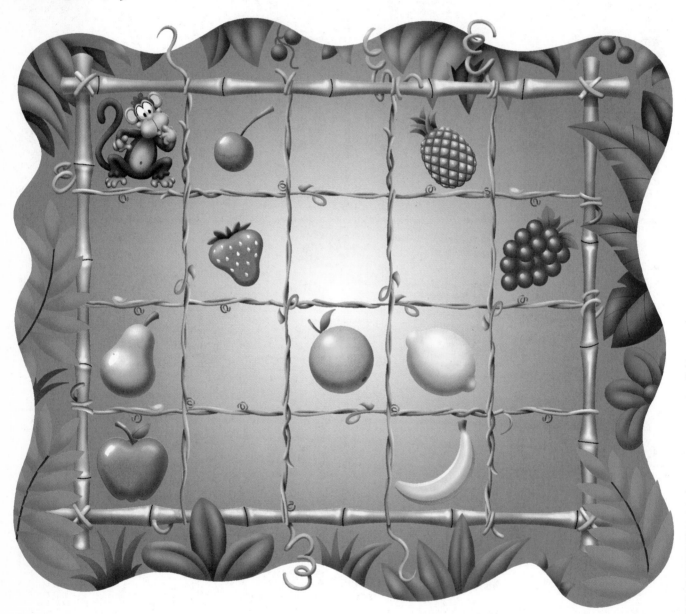

Directions:
Mr. Monkey is hungry! Follow the directions to move Mr. Monkey to his dinner.

1. Move Mr. Monkey to 1 space to the right of the cherry. Draw a triangle in this space.

2. Move Mr. Monkey 2 spaces to the right of your triangle. Draw a circle in this space.

3. Move Mr. Monkey 3 spaces below your circle. Draw a square in this space.

4. Move Mr. Monkey 1 space to the left of your square.

5. Circle the food Mr. Monkey wants to eat.

Name _____

Front or Back

①

②

③

④

⑤

Directions:
1. Circle the picture of the front of the monkey.
2. Circle the picture of the back of the duck.
3. Draw an X on the back of the bunny.
4. Draw an X on the front of the giraffe.
5. Circle the pictures that show the back of a bear.

Name

Inside or Outside

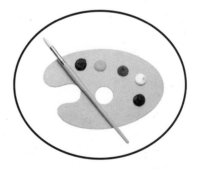

Directions:

1. Draw an X outside the house.
2. Draw an X inside the basket.
3. Draw an X on the toy car outside the circle.
4. Draw an X on the paint set inside the circle.

Name _____

Solve Puzzles

1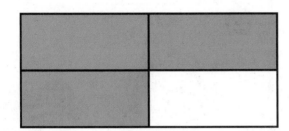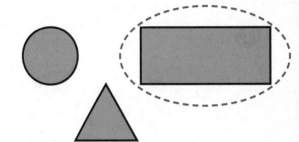

2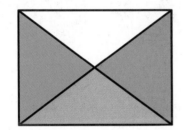

3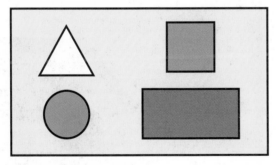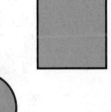

4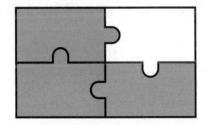

Directions:
1–4. Circle the object that fits in the blank space.

Chapter 7   Lesson 7

Name _____

Progress Check 2

 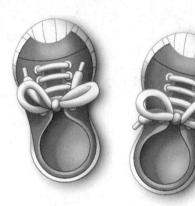

Directions:
1. Circle the sock on the left.
2. Circle the shoe on the right.
3. Draw an X on the front of the toy.
4. Draw an X on the object inside the doghouse.

100 one hundred

Name _____

Review

①

②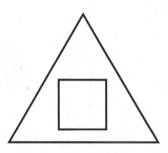

③

④

⑤

⑥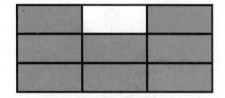

Directions:

1. Circle the bird before the yellow bird.
2. Draw an X on the shape inside the triangle.
3. Circle the kitten in the middle.
4. Circle the dog on the right.
5. Circle the dog below the dog bowl.
6. Circle the object that fits in the blank space.

Name _____

Test

1

2

3

4

5

6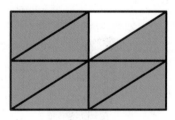

Directions:

1. Circle the flower after the red flower.
2. Draw an X on the shape inside the square.
3. Draw an X over the plant in the middle.
4. Circle the flower on the left.
5. Circle the leaf above the tree.
6. Circle the object that fits in the blank space.

Sort Objects by Attributes

Home Connection

English

Spanish

Dear Family,
Today I started in **Chapter 8, Sort Objects by Attributes.** I will learn to sort objects by their characteristics. I will describe objects using words such as "tall" and "short" or "full" and "empty."

Love, _____

Querida Familia:
Hoy empecé el **Capítulo 8, Agrupar objetos por atributos.** Aprenderé a agrupar objetos por sus características. Describiré objetos usando palabras como "alto" y "corto" o "lleno" y "vacío".

Con cariño, _____

Help at Home

Work with your child to sort objects by their characteristics. Fill two cups with different amounts of cereal and have your child describe each cup. Is it full or empty?

Ayuda en casa

Ayude a su hijo a agrupar objetos de acuerdo a sus características. Llene dos tazas de diversas cantidades de cereal y haga que su niño describa cada taza. ¿Es llena o vacia?

Full   Empty

Math Online Take the chapter Quick Check quiz at macmillanmh.com.

Name

Get Ready

1

2

3

4

Directions:
1. Look at the first shape. Circle the shape that matches.
2. Look at the heart. Circle the heart that is smaller.
3. Look at the star. Circle the star that is bigger.
4. Circle the shapes that have curved sides. Draw an X on the shapes that have straight sides.

Name _____

Same or Different

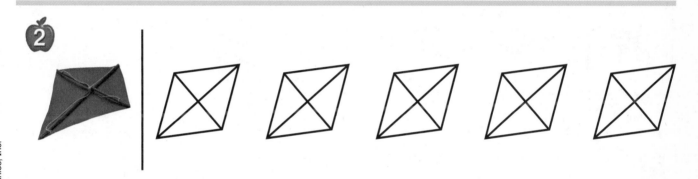

Directions:

1. Look at the first umbrella. Circle all of the umbrellas that are the same color. Draw an X on the umbrellas that are a different color.

2. Look at the first kite. Color three kites the same color. Color the other kites a different color than green.

3. Look at the first balloon. Circle the balloons that are the same size as the first balloon. Draw an X on the balloons that are a different size.

Name _____

Equal or Unequal

 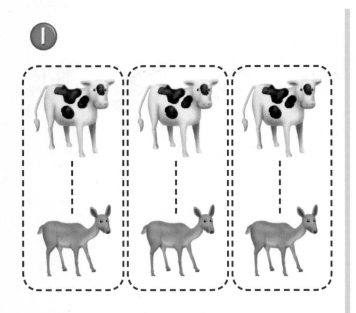

Directions:

1. Draw lines to match the cows and the deer. Circle them if they are equal or draw an X on them if they are unequal.
2. Draw lines to match the horses and the goat. Circle them if they are equal or draw an X on them if they are unequal.
3. Draw lines to match the pigs and the rabbits. Circle them if they are equal or draw an X on them if they are unequal.
4. Draw lines to match the ducks and the cats. Circle them if they are equal or draw an X on them if they are unequal.

Name _____

More or Less

①

②

Apple
Juice

3

Directions:

1. The cup holds 5 marbles. The pitcher holds 50 marbles. Draw an X on the container that holds less.

2. The carton has enough milk for 10 people. The juice box has enough juice for one person. Circle the container that holds more.

3. Sally is playing in the sand at the beach. She is using the bucket shown. Draw a bucket that will hold more sand than Sally's bucket.

Name _____

Long or Short

1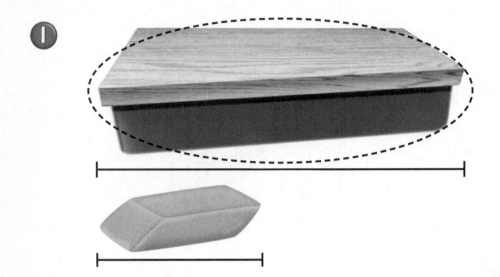

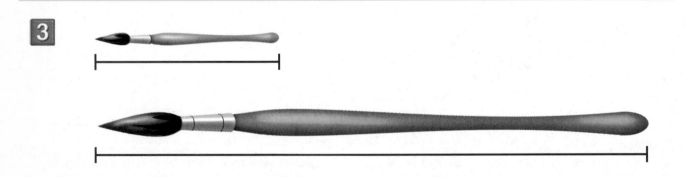

2

3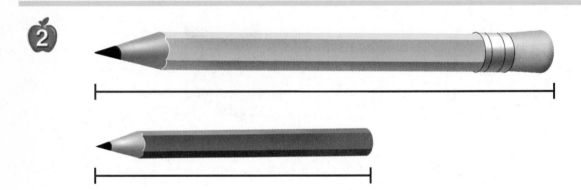

Directions:
1. Circle the long eraser.
2. Draw an X on the short pencil.

3. Circle the long paintbrush.

108 one hundred eight

Name _____

Progress Check 1

_____   _____   Yes   No

 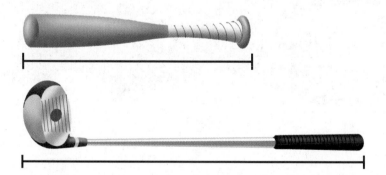

Directions:

1. Look at the first ball. Circle the balls that are the same size. Draw an X on the balls that are a different size.
2. Count the fish. Write how many. Count the frogs. Write how many. Is the number of fish equal to the number of frogs? Circle Yes or No.
3. Corey and Matt went to the beach. Corey counted 3 starfish and Matt counted 4 shells. Circle the group that has more. Draw an X on the group that has less.
4. Draw an X on the short object.

Name _____

Replay

Directions:

Tamika and Evan are having lunch. Follow the steps to complete the picture.

Step 1: Color Tamika's hair a different color from Evan's hair.

Step 2: Color both of their shirts the same color.

Step 3: Evan and Tamika have an equal number of apples. Draw Evan's apples.

Step 4: Evan has more carrots than Tamika. Draw Evan's carrots.

Step 5: Color the cup orange that holds more juice.

Name _____

Tall or Short

1

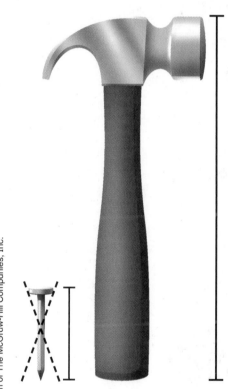

2

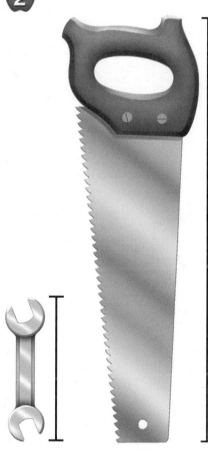

3

Directions:

1. Draw an X on the short object.
2. Circle the tall object.

3. Draw an X on the short object.

Name

Heavy or Light

1

2

3

4

Directions:
1–4. Circle the heavy object. Draw an X on the light object.

112 **one hundred twelve**

Name _____

Full or Empty

 1

 2

3

 4

Directions:

1–4. Circle the full items. Draw an X on the empty items.

Chapter 8   Lesson 7   **one hundred thirteen 113**

Name _____

Progress Check 2

①

②

③

Directions:
1. Look at the animals. Circle the animal that is tall.
2. Look at the animals. Draw an X on the animal that is heavy.
3. Look at the dogs and their bowls. Circle the dog that has an empty bowl.

Name _____

Review

①

②

③

④ |

Directions:
1. Circle the tall tree.
2. Circle the object that is heavy.
3. Circle the vine that is long. Draw an X on the vine that is short.

4. Look at the first flower. Circle the flowers that are the same as the first flower. Draw an X on the flowers that are different from the first flower.

Name _____

Test

①

②

③

_____

_ _ _ _ _ _ _

_____

_____

_ _ _ _ _ _ _

_____

Yes   No

④

Directions:

1. Circle the object that is tall. Draw an X on the object that is short.
2. Circle the popcorn box that is full. Draw an X on the popcorn box that is empty.
3. Count the piñatas. Write how many. Count the pinwheels. Write how many. Is the number of piñatas equal to the number of pinwheels? Circle Yes or No.
4. Look at the cup. Color two cups the same color. Color the other cups a different color.

Home Connection

English

Dear Family,
Today I started **Chapter 9, Order Objects by Attributes.** I will learn to compare and order objects by describing their physical characteristics, such as length, height, and weight.

Love, _____

Spanish

Querida Familia,
Hoy empecé **Capítulo 9, Ordenar objetos por sus características.** Aprenderé a comparar y ordenar objetos describiendo sus características físicas, como longitud, altura, y peso.

Con cariño, _____

Help at Home

Help your child order objects by attributes. At home, place three different objects together. Have your child order them based on physical characteristics, such as smallest to largest or longest to shortest.

Math Online > Take the chapter Quick Check quiz at macmillanmh.com.

Ayuda en casa

Ayude a su hijo a ordenar objetos de acuerdo a sus características. Busque tres objetos diferentes en su casa y júntelos. Pídale a su hijo que ordene los tres objetos basándose en sus características físicas, como de más pequeño a más grande o de más largo a más corto.

Name _____

Get Ready

 1

 2

3

 4

Directions:
1. Circle the bigger plant.
2. Circle the smaller watering can.

3. Circle the bigger flower pot.
4. Circle the smaller flower.

Name

Long, Longer, Longest

①

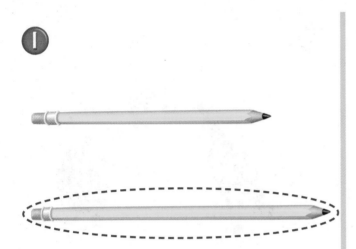

②

3

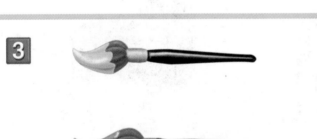

④

Directions:
1. Circle the pencil that is longer.
2. Circle the piece of chalk that is longer.

3–4. Circle the longest object.

Name _____

Tall, Taller, Tallest

1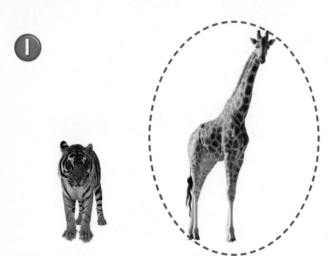

2

3

4

Directions:

1–2. Circle the animal that is taller.
  3. Circle the tallest animal.

4. Circle the group that is in order from tallest to shortest.

Name

Short, Shorter, Shortest

①

②

③

④

Directions:
1–2. Circle the shorter animal.
   3. Circle the shortest animal.

4. Circle the group that is in order from shortest to tallest.

Chapter 9   Lesson 3   **one hundred twenty-one**   **121**

Name _____

Heavy and Heavier

1

2

3

4

Directions:
1–2. Circle the heavier object.
  3. Circle the object that is heavier than the eraser.
  4. Circle the object that is heavier than the notebook.

Name _____

Progress Check 1

①

②

3

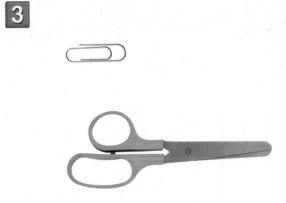

④

Directions:
1. Circle the group that is in order from shortest to longest.
2. Circle the taller plant.

3. Circle the shorter object.
4. Circle the toy that is heavier than a football.

Name

Replay

Directions:

Oh, no! There is a problem at the zoo! Last night, the zookeeper forgot to lock some of the cages and now the animals are free. Help the zookeeper get the animals back in their cages.

Step 1: Draw a line from the shortest animal to the shortest cage.

Step 2: Draw a line from the tallest animal to the tallest cage.

Step 3: Draw a line from the longest animal to the longest cage.

Step 4: Draw a line from the heaviest animal to the last cage.

Name

Light and Lighter

 1

 2

3

4

Directions:
1–2. Circle the fruit that is lighter.
  3. Circle the fruit that is lighter than the bananas.

4. Circle the fruit that is lighter than the pineapple.

Name

More and Most

1

2

3

4

Directions:

1–2. Circle the object that holds more.

3–4. Circle the object that holds the most.

Name _____

Less and Least

1

2

3

4

Directions:
1–2. Circle the item that holds less.

3–4. Circle the item that holds the least.

Progress Check 2

①

②    **③**

④    **⑤**

Directions:
1. Circle the object that is lighter than the pot.
2. Circle the object that holds the least.
3. Circle the object that holds less.
4. Circle the object that holds the most.
5. Circle the cup that holds more.

Chapter 9   Progress Check

Name _____

Review

1

2

3

4

5

Directions:
1. Circle the animal with the longer tail.
2. Circle the shortest animal.
3. Circle the tallest animal.
4. Circle the animal that is heavier than the rabbit. Draw an X on the animal that is lighter than the rabbit.
5. Circle the object that holds the most. Draw an X on the object that holds the least.

Name _____

Test

 1

 2

3

 4

Directions:
1. Circle the longest object. Draw an X on the shortest object.
2. Circle the tallest animal. Draw an X on the shortest animal.
3. Circle the fruit that is heavier than the pineapple.
4. Circle the object that holds the most. Draw an X on the object that holds the least.

130 **one hundred thirty**

Home Connection

English   **Spanish**

Dear Family,
Today I started **Chapter 10, Patterns.** I will learn how to identify and extend simple patterns related to sight, sound, and texture.

Love, _____

Querida Familia:
Hoy empecé el **Capítulo 10, Patrones.** Aprenderé cómo identificar y ampliar patrones simples relacionados con la vista, el sonido y el tacto.

Con cariño, _____

Help at Home

Help your child identify and extend simple patterns. Clap, slap your knees, and hop on one foot in a pattern for your child to copy. Have your child make up a different pattern.

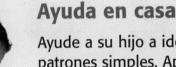 Take the chapter Quick Check quiz at macmillanmh.com.

Ayuda en casa

Ayude a su hijo a identificar y ampliar patrones simples. Aplauda, golpee sus rodillas y golpee con sus pies siguiendo un patrón ABCABC para que su hijo lo copie. Pídale a su hijo que él cree su patrón diferente.

Name _____

Get Ready

1

2

3 1 2 3 4 _____

4 1 ___ 3 4 5 6

5 1 2 3 ___ 5 6

Directions:

1–2. Draw an X on the figure that is different.

3–5. Write the missing numbers.

Name _____

More than One Attribute

①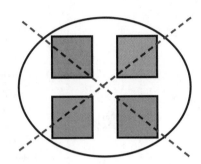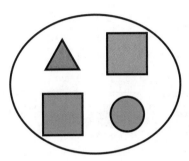

②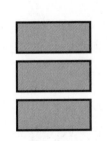

③

④

Directions:
1. Draw an X on the group that shows figures of the same size and shape.
2. Circle the figure that belongs in the group.
3. Draw and color a figure that is the same size and shape as those shown.
4. Circle the triangles. Draw an X on the rectangles.

Chapter 10   Lesson 1

one hundred thirty-three   **133**

Name

AB Patterns

1

2

3

Directions:
1. Circle the insect that comes next in the pattern.
2. Circle the insect that comes next in the pattern.

3. Draw and color the butterfly that comes next.

Name _____

AAB Patterns

①

   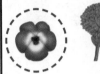

②

③

Directions:
1. Circle the flower that comes next.
2. Circle the plant that comes next.
3. Draw and color the leaf that comes next.

Name _____

ABB Patterns

1

2

3

4

Directions:
1–3. Circle the fruit that comes next.

4. Draw the fruit that comes next.

Name _____

Progress Check 1

 ❶

❷   |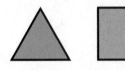

❸ ★ ★ ☾ ☆ ★ ★ ☾ ☆ ★ ★ ☾ | ☾ ☆

🐟❹  |

Directions:
1. Circle the squares. Draw an X on the triangles.

2–4. Circle the figure that comes next in the pattern.

Chapter 10   Progress Check

one hundred thirty-seven **137**

Copyright © Macmillan/McGraw-Hill • Glencoe, a division of The McGraw-Hill Companies, Inc.

Name _____

Replay

Directions:
The squirrel collects vegetables. Draw a line from each row to show which vegetable comes next.

Name _____

ABC Patterns

①

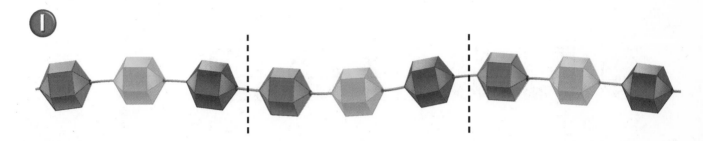

②

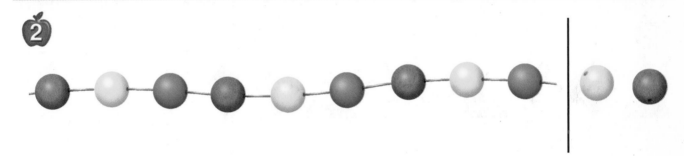

③

④

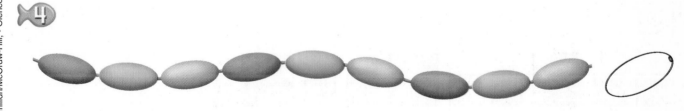

Directions:

1. Draw a line to show where the pattern starts over, or repeats.

2–3. Circle the figure that comes next in the pattern.
4. Color the white bead to show what comes next.

Name _____

Identify and Extend Patterns

②

③

④

Directions:
1–4. Look at the pattern. Draw and color the next figure.

Name _____

Create Patterns

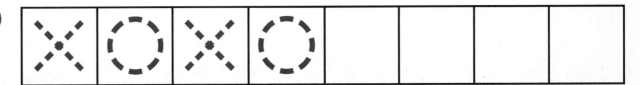

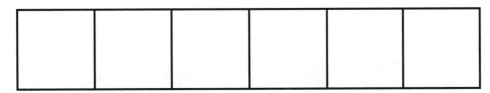

Directions:
1. Use the letters x and o to make a pattern.
2. Choose two numbers. Make an ABB pattern.
3. Choose three figures. Make an ABC pattern.

4. Choose a letter, a number, and a figure. Draw a pattern in the boxes. Use one letter, number, or figure in each box.

Name _____

Progress Check 2

①

②

③

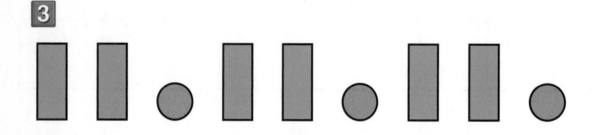

④

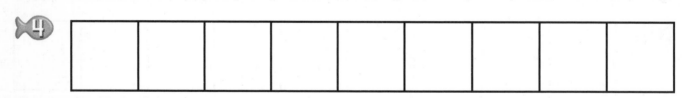

Directions:

1. Circle the fruit that comes next.
2. Look for a pattern. Color the apple to show which color comes next.

3. Look for a pattern. Draw and color the next figure.
4. Choose two figures. Make a pattern.

Name _____

Review

1 |

2

3 |

4   |

Directions:
1. Circle the insect that comes comes next in the pattern.
2. Circle the figure that comes next.
3. Circle the bird that comes next.
4. Draw the figure that comes next.

Chapter 10   Review   **one hundred forty-three**   **143**

Name _____

Test

1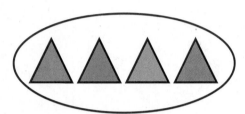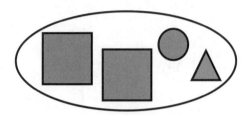

2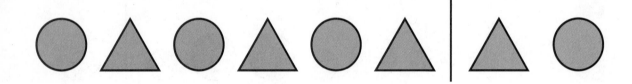

3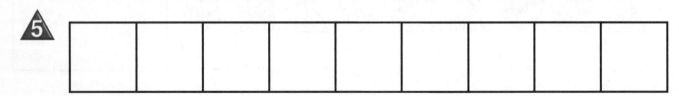

4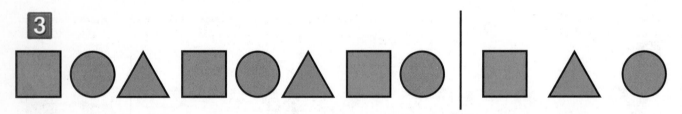

5

Directions:

1. Draw an X on the group that shows the same size and shape.

2–4. Circle the figure that comes next.
5. Choose two figures. Make a pattern.